Prepare yourself
for a creative journey!
If you color in the book
be sure to place a sheet of card stock
behind the page to protect the next page
from bleeding pens or markers.

Thank you to the following amazing women for allowing me to use their colorings of my designs:

Antoniette Watson Agee

Rosa Lee Ramsay

Kat Suydam

Shela Walton

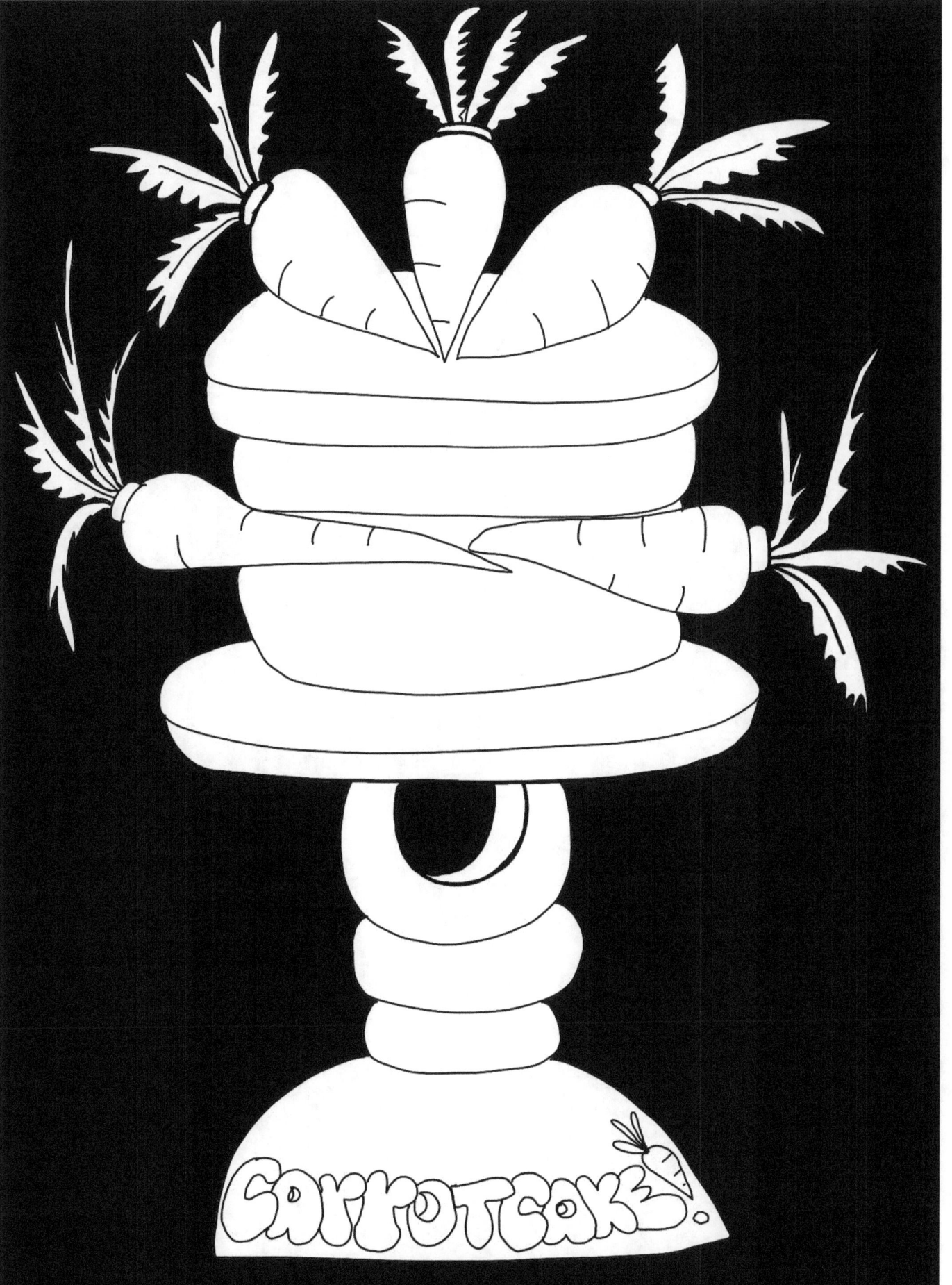

d. mcdonald designs

Cakes, Castles, & Outer Space

White & Black Backgrounds Two

d. mcdonald designs
Cakes, Castles, & Outer Space
White & Black Backgrounds Two

D. McDonald Designs Adult Coloring Buffet

D. McDonald's Mysteria Full Page Size by ms deborah mcdonald

d. mcdorald designs
Cakes, Castles & Outer Space
White & Black Backgrounds Edition

d. mcdonald designs
Fabulous Florals Two

d. mcdonald designs
Butterflies, Bugs and Black Backgrounds

Other Books
by
Deborah L. McDonald

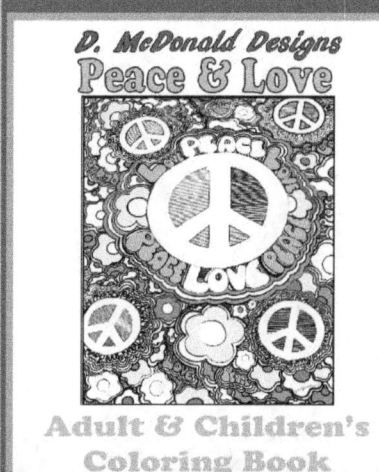

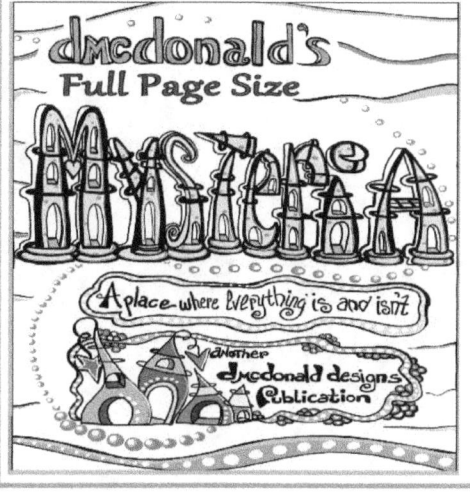

D. Mc Donald Designs

Toddler
Time
Color & Learn
Series

Book One

Featuring No Fail
Toddler Friendly **Black** Backgrounds
Learning Can Be Fun !

D. McDonald Designs

Butterflies,
Bugs,
and
Black
Backgrounds
A Family Fun Coloring Book

d.mcdonald designs
Cakes,
Castles,
&
Outer
Space
White & Black
Backgrounds Edition
Adult & Children's
Coloring Book

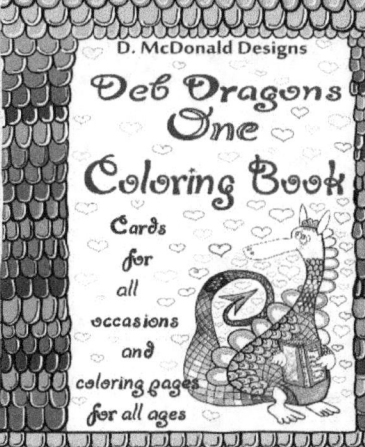

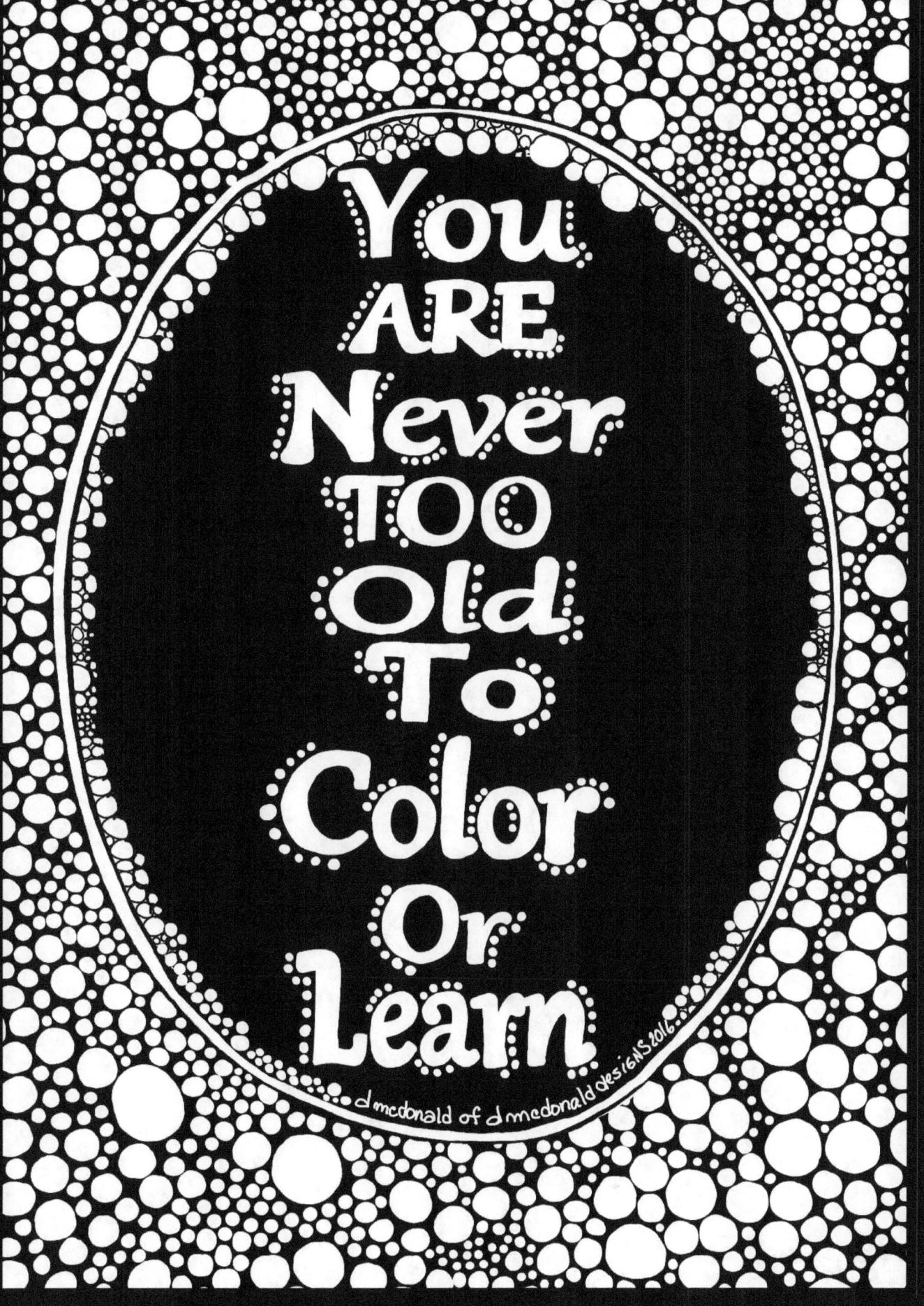